# China's Tibetan Mastiff

The Tibetan Mastiff lives in Tibet,
the most mysterious snowy plateau in the world.
Boasting ferocity rivaling that of lions and tigers,
aristocratic aloofness and elegance,
unquestionable loyalty,
and an amazingly high degree of intelligence,
the Tibetan Mastiff fully deserves
the honor of being titled "the Holy Dog."

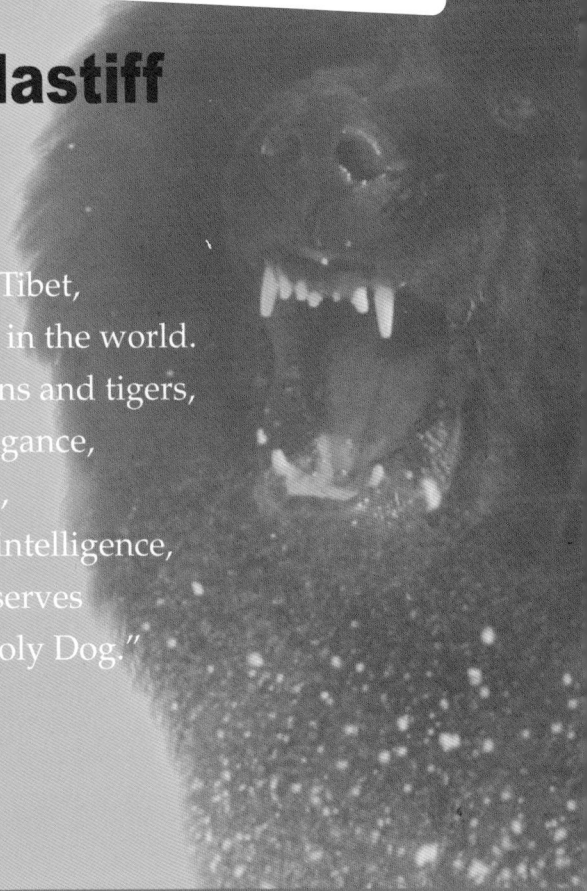

First Edition 2006

**China's Tibetan Mastiff**

ISBN 7-119-04272-6

© Foreign Languages Press
Published by Foreign Languages Press
24 Baiwanzhuang Road, Beijing 100037, China
Home Page: http://www.flp.com.cn
E-mail Addresses: info@flp.com.cn
                  sales@flp.com.cn
Distributed by China International Book Trading Corporation
35 Chegongzhuang Xilu, Beijing 100044, China
P.O. Box 399, Beijing, China

*Printed in the People's Republic of China*

# China's Tibetan Mastiff

**The Tibetan Mastiff is a "living fossil" for the study of the origin and evolution of dogs**

From living in the wild to being domesticated by human beings, the Tibetan Mastiff has accompanied us for centuries. But it retains many of the characteristics of primitive dogs from appearance to living habits.

**The Tibetan Mastiff deserves the honor of being titled "world stud dog"**

From Central Asia to Europe, Tibetan Mastiffs have mixed with other dog species along with the expansion of human activities, and have become the ancestors of many modern species of giant dogs.

Tibetan Mastiff - King of the Plateau

**The Tibetan Mastiff can rival lions and tigers with its ferocity**

Besides its giant size, people are alarmed at its fierce visage and thunderous bark. The Chinese have a saying that one Tibetan Mastiff can put three wolves to flight, and two Tibetan Mastiffs can subdue a tiger.

**The Tibetan Mastiff is an elegant aristocrat**

Like human nobles born into renowned European families with hundreds of years of history, Tibetan Mastiffs may also feel proud of the blue blood running in their veins, and therefore often hold themselves aloof from other breeds. You will never find a Tibetan Mastiff being greedy, wretched or fawning.

The beautiful Shangri-La Grassland

### The loyalty of the Tibetan Mastiff is unquestionable

The Tibetan Mastiff regards itself as one of the family members, rather than as a pet. To protect the family property and the lives of its family is all it lives for. Its concern extends even as far as the family's cattle and horses.

### The Tibetan Mastiff is a born fighter

Whoever has seen a fight between a Tibetan Mastiff and its enemy will have such an impression. Its instinct is to forge ahead rather than fall back, to attack rather than evade, to bleed rather than yield.

### The Tibetan Mastiff has a high degree of intelligence

It is an animal that for centuries has maintained its own way of thinking, and the habit of making decisions on its own. It does not simply obey its master's orders; it can also make judgments that benefit its master, based on its own observation of the situation. No one knows where it gets such an amazing ability.

Mt. Minling

Be careful! You have entered its territory.

# Preface

   "Ao" (mastiff) is a word the Chinese have used for a giant ferocious dog since ancient times. According to the *Er Ya* (*Near to Correctness*), the earliest Chinese dictionary, compiled in the second century BC and one of China's ancient classics, "Ao is a dog that is longer than four *chi* (three *chi* equals approximately one meter)." The *Shuo Wen Jie Zi* (*An Analytical Dictionary of Chinese Characters*), another important Chinese dictionary, explains "Ao" as "unrestrained," which may be interpreted in modern Chinese as "wild, fierce, obstinate and unruly."

   In fact, most of China's giant fierce dogs live in the west and north, or the main pasture areas in Qinghai and Gansu provinces, the Tibet, Inner Mongolia and Xinjiang Uygur autonomous regions, western and northern Sichuan Province and part of Yunnan Province.

View of the Tibetan grassland

Despite the poor natural conditions there, a dozen hardy ethnic minorities have dwelt in the northwestern region of China for thousands of years. They relied on hunting and pasturing, and bred giant dogs as assistants in their life and production, and even adopted them as part of their families and loyal guardians.

A herder of Tibet's Khampa people and his loyal "comrade-in-arms" — a Tibetan Mastiff

བོད་ཁྱི།

The Tibetan Mastiff is an outstanding representative of such dogs. The breed mainly lives in the frigid zone at an altitude of 3,000-5,000 m on the Qinghai-Tibet Plateau, and are domesticated by the Tibetan people, hence the name. It has also had other Chinese names, such as Black Lion-Dog, Songpan Dog and Tibetan Dog. The British Kennel Club named it the "Tibetan Mastiff," while the Tibetans usually call it "Dokhyi" (chained dog), as they are usually chained to a pole on the grazing grounds. They will bark and leap wildly when strangers approach them, causing the chains to rattle.

Unfortunately, the number of purebred Tibetan Mastiffs is decreasing. The animal has been listed in Category II under state protection.

Dokhyi, the Tibetan name for the Tibetan Mastiff, means a tethered dog. No stranger can approach a Tibetan Mastiff or the territory it guards if the dog is not tethered by an iron chain to a wooden stake.

The terrifying snarl of a Tibetan Mastiff

The Tibetan Mastiff is a lover of freedom.

# Contents

བོད་ཁྲི།

# Legends and Origin of the Tibetan Mastiff

*Religion and Reality*

བོད་ཁྱི།

A nation with a long history always has numerous beautiful and moving stories. The Tibetan Mastiff, the companion and guardian of man, is the subject of many touching legends. One of them goes like this: Long, long ago, one severe winter the land of the Tibetans was frozen over, and plagues prevailed everywhere. The people and the livestock they relied on suffered hunger, cold and illness. All of a sudden, a large number of living Buddhas wearing cloaks, holding

Typical landscape of the Shangri-La Grassland

sacred bells and riding on giant beasts (which were later found to be Tibetan Mastiffs) descended from Heaven. With the arrival of the living Buddhas and the Tibetan Mastiffs, the ice melted, the plagues were dispelled, the land resumed its vitality, and the people were saved. Ever since then, the Tibetan people have held Tibetan Mastiffs in high esteem, and believe they are envoys from Heaven, divine dogs with a mission to protect the herdsmen.

Legend has it that the Tibetan Mastiff was the steed of a living Buddha.

Both Buddhist pagodas and Tibetan Mastiffs are held to be sacred by Tibetans.

The Tibetan region of China has a long history of reverence for Buddhism, and that's why many of the local legends are related to religion. Here's another story about the origin of the Tibetan Mastiff.

1. Turn the prayer wheel clockwise and one spin of the wheel amounts to saying the Six-Word Mantra once.

2. Dancing sutra streamers inscribed with the Six-Word Mantra or other sutra texts, are hung on roofs, hilltops and "mani" mound, to transmit prayers in the wind.

3. Sutra streamers and a Tibetan Mastiff

4,5,6. A "mani" stone mound with the Six-Word Mantra carved upon it. According to the tenets of Tibetan Buddhism, a "mani" acts as a prayer altar which also helps to fend off evil spirits.

3

4

5

6

A long time ago, a Tibetan hunter was attacked by a leopard in remote mountains. The man had no time even to draw his bow, but right at that moment, a giant dog sprang down from Heaven, and stood between the leopard and his intended victim. With a thunderous roar, the dog attacked the leopard. In a welter of blood, the dog grappled with the beast, giving the hunter a chance to dispatch the leopard with his sword. In deep gratitude, the hunter took the dog home with him; ever afterwards the dog became the hunter's constant companion, and guarded his herds and homestead.

བོད་ཁྱི།

It's not important whether the story is true or not. It describes how the origin and development of dogs, the loyal companions of man, were closely related to human activities. But now let's take a look at the origin of the Tibetan people.

The ancestors of the Tibetan people lived along the middle reaches of the Yarlung Zangbo River. Like many other Stone Age people, they lived in hunter-gatherer

groups before settling down to farm the land. Early on, some Tibetans migrated to the north and east, and mixed with other nomadic tribes such as the Xirong and Qiang. During the long course of their historical development, they spread into what are today Sichuan, Qinghai and Gansu provinces, and evolved into the Tibetan people of today.

The Garze Tibetan Autonomous Prefecture, Sichuan Province, is commonly known as the Kham Area.

The development process of the Tibetan people was one of adaptation to severe natural conditions. On the Qinghai-Tibet Plateau, with an average altitude of over 3,000 m, the Tibetans settled down in places where there was water and grass. Though they did some farming, hunting and pasturing still dominated their life. During the process, the herdsmen had to adapt to the harsh climate and complicated geographical conditions, and keep vigilance against wolves, bears and leopards that threatened their own lives and those of their flocks. This made it highly urgent to domesticate dogs that were brave and aggressive to guard against the predators.

The skeleton of a yak molders on the grassland. The Tibetan prairie is a beautiful but harsh place.

Although herding is the main occupation of Tibetans, they do some farming too.

To survive the cruel conditions, some primitive dog breeds became bigger, tougher, fiercer and cleverer, in accordance with the law of survival of the fittest. When they were short of food, they would follow the herdsmen in the hope of feeding on cattle which had died of illness, and other offal. Soon, the herdsmen began to domesticate them, and the dogs learned to guard the nomadic camps and livestock herds, and became irreplaceable "members" of the families of the herdsmen. These were the original Tibetan Mastiffs.

Over the centuries, Tibetan Mastiffs became like family members to the Tibetans.

བོད་ཁྲི།

# History of the Tibetan Mastiff

*Ancient Records*

The Tibetan Mastiff is one of the most ancient and rarest of all the working breeds in the world. They roamed the Qinghai-Tibet Plateau as early as 5,000 years ago.

The earliest Chinese record of a Tibetan Mastiff dates from the 11th century BC. According to this account, King Wu, who founded the Zhou Dynasty (c.1100-771BC), received a giant dog as a gift from the headman of a tribe living in western China. The dog was four *chi* high, and had been trained to hunt humans of various skin colors. The dog is thought to have been an ancient Tibetan Mastiff.

Old pictures of Tibetan Mastiffs

Old pictures of Tibetan Mastiffs

The *Spring and Autumn Annals*, which record China's history from 722 to 481 BC, mentions in the "Story of Gongyang" that a certain prince released a mastiff, which he had been given as tribute from a western tribe, and urged it to attack an official. But the dog was killed by the latter's bodyguard. The official sneered at the prince: "Your mastiff is inferior to mine."

Later, Tibetan Mastiffs were trained to accompany armies. Genghis Khan (1162-1227), founder of the vast Mongol empire, had a contingent of 30,000 Tibetan Mastiffs raised in the Tibetan area. The dogs were cited for their contribution to the establishment of the Yuan Dynasty (1271-1368). Later, Tibetan Mastiffs found their way via Central Asia and Europe and became the ancestors of improved local breeds. For instance, the Saint Bernard, Great Pyrenees, Mastiff, Great Dane and Caucasian Shepherd Dog are all descended from Tibetan Mastiffs, which is why the Tibetan Mastiff is known as the "world stud dog."

Tibetan Mastiffs guard the herders' tents and livestock.

There are many records about Tibetan Mastiffs in other parts of the world, but most of them are inaccurate. One was left by Ktesias, a Greek royal doctor in the 5th century BC. He described a mytho-logical animal which lived on high peaks and looked half bird and half dog, with reddish-brown chest and feet. This description is quite close to the appearance of the Tibetan Mastiffs. Aristotle (384-322 BC), a famous Greek philosopher and scientist, also mentioned an "Indian dog," which was huge and robust, with a big head and a wide mouth. Obviously, this description accords less with the Indian breeds, and more with the features of the Tibetan Mastiff.

A more accurate description of the Tibetan Mas-tiff came from Marco Polo, the Italian traveler who visited China in the 13th century. In *The Travels of Marco Polo*, he describes a kind of giant and aristo-cratic dog in China which was "very good at pull-

The Tibetan Mastiff is not only a treasure of the Tibetan nationality, but has long since won favor from people in other places and of other nationalities in China.

Such vigorous and energetic pedigree Tibetan Mastiffs can rarely be seen these days.

བོད་ཁྱི།

ing down game," especially musk deer. Called by the Tibetans "Ao," they were "as big as donkeys." He says that their lips were raised upward and there were deep wrinkles around their eyes.

It is said that Tibetan Mastiffs were exported to Greece some 2,000 years ago, then to the Roman Empire, and further to the Slav countries. To date, many world-renowned dog breeds still bear the genes of Tibetan Mastiffs, and there is evidence that in the 18th and 19th centuries, many Europeans knew about them. In 1847, the Indian governor-general presented a "Tibetan dog" (Tibetan Mastiff) to Queen Victoria. Today, the ancient and mysterious breed is gaining increasing favor from people in the US, Europe, South Korea and Japan.

This Tibetan Mastiff retains a great deal of the facial and head features of the ancient pure-blood Tibetan Mastiffs.

བོད་ཁི།

# Characteristics of the Tibetan Mastiff

*Unveiling the Mysteries*

Its unique living conditions give the Tibetan Mastiff its unique features. A single look is enough to recognize its ferocity. The breed is divided into two types by their appearance: lion and tiger. So, the belief of many, including Aristotle, who thought that the Tibetan Mastiff had some blood lineage with lions or tigers was not without foundation.

Lion-type Tibetan Mastiff

Lion-type Tibetan Mastiff

The differences are mainly in the head. The lion-like Tibetan Mastiff has long and thick hair, which flares outward. Since its head is already very big, this makes it very much like that of a lion. The tiger-like Tibetan Mastiff has short and stiff hair, and a shorter but wider mouth. The lion-like ones look larger when their long hair sticks up, but in fact the tiger-like ones are stronger and more valiant.

No one dares to look directly into the eyes of a strange Tibetan Mastiff. The somber, deep and cold mystery in its eyes often makes people tremble with fear. If it stares at you, you'll be tempted to turn and flee. The eyes of a Tibetan Mastiff are nearly triangular, tilting downward from the upper corners. The color ranges from pale to dark amber. Some have two white or yellow dots above the eye sockets. Those with a black back and brown belly, nicknamed "tiebaojin" or "black and tan," are said to be the offspring of the "Yelling Heavenly Dog" raised by Erlangshen, a deity in the Chinese mythological novel *Journey to the West*. Legend says that the divine dog had two additional eyes that could see things 1,000 *li* away.

Black and tan Tibetan Mastiff

བོད་ཁྱི།

The eyes of Tibetan Mastiffs are not sensitive to most colors, except bright red and yellow. So, you should not appear before them in these two colors, and be sure never try to irritate them as the Spanish matador does in a bullfight.

Despite its commanding appearance, the Tibetan Mastiff is not the biggest of all dogs. A modern one is usually 60-90 cm high, with few over 90 cm high, and the body length is the same or a little longer. Nevertheless, the figures are enough to make the dog one of the giant breeds. For a giant breed, a well-balanced stature is a key element. A muscular neck, an ideal angle between the shoulder and forelegs (with shoulder breadth equaling leg length), flat and wide hips, and thick and tenacious toe callus, all equip a Tibetan Mastiff with enormous strength and high flexibility, and enable it to launch agile attacks.

བོད་ཁྱི།

What makes the Tibetan Mastiff unique is its double layer of hair, which no other dogs have. The undercoat is short and soft, while the outer hair is coarse. They form a solid armor to protect the dog from the frigid, windy and dry conditions on the plateau. The hair of pedigree Tibetan Mastiffs is pure black, black and tan, yellow, brown, or pure white. Those with other colors are inferior breeds. Under the serfdom

Tibetan Mastiff with a golden coat

system practiced in Tibet before its liberation by the communists in the 1950s, pedigree Tibetan Mastiffs were the exclusive property of the privileged class, which included Living Buddhas, lamas, tribal chiefs and serf owners. Those with reddish-brown hair were raised only in monasteries, as their color was similar to that of the lama cloaks and they were therefore deemed as divine by the lamas. Ordinary Tibetans were prohibited from raising the dogs.

A typical Tibetan Mastiff with a black coat

Black and tan Tibetan Mastiff

Snow-white Tibetan Mastiff

བོད་ཁྱི།

# Breeds of the Tibetan Mastiff

*Like Attracts Like*

བོད་ཁྱི།

Over millennia of co-existence and development in the company of human beings, Tibetan Mastiffs have expanded to almost all parts of the world, and they have produced new hybrid breeds with other dogs. However, the number of superior purebred Tibetan Mastiffs is not increasing, but decreasing.

To help the breed multiply, many people from all over China and other parts of the world have endeavored to search for fine-breed Tibetan Mastiffs for artificial propagation. The dogs thus bred are from three major lines: Lhasa, Yushu and Hequ.

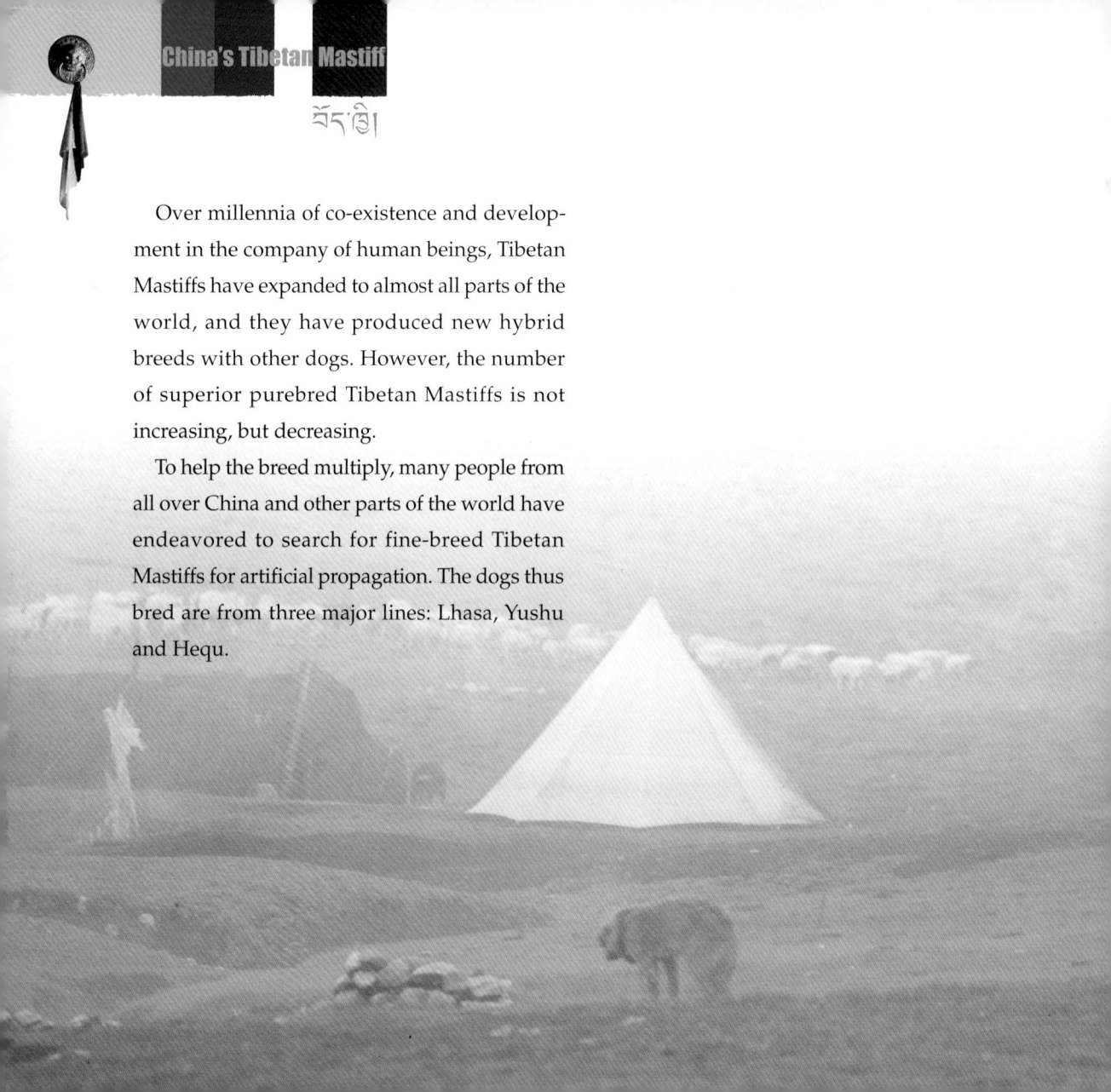

The Lhasa-line refers to Tibetan Mastiffs that live in the frigid area around Lhasa, capital of the Autonomous Region, and covering Ngari, Zetang, Nagqu counties, and northern Tibet. Since many other dog breeds also live in these areas, and the herdsmen are indifferent to the breeding of inferior strains, cross-breeding is common. But in northern Tibet, fine qualities of Lhasa-line Tibetan Mastiffs have been maintained due to the relative seclusion of the region and the sparse habitations of the herdsmen. These dogs are huge and strong, and they have short hair, a big and square head, and sturdy legs. The Lhasa-line Tibetan Mastiffs are mainly tiger-like, and most are black on the back and brown on the belly.

Tiger-type Tibetan Mastiffs in Lhasa

The Yushu-line refers to Tibetan Mastiffs that live in an area centered on Yushu in Tibet, and in Golog, Haixi, Hainan and Ma Qu Huang He in neighboring Qinghai Province. The climate in Qinghai is bitterly cold in winter, and cool and humid in summer.

Scene of the Yushu Area

བོད་ཁྱི།

Tibetan Mastiff in
the Yushu Area

The Tibetan Mastiffs living there are lion-like, short, with a small head and a narrow face, but thick coats. The Yushu-line Tibetan Mastiffs come in various colors, the pure white ones being called "xue-ao" (snow mastiff). The latter are extremely rare and valuable.

Snow-white Tibetan Mastiff

The best of all Tibetan Mastiffs come from Hequ—the first bend on the Yellow River, in southeast Qinghai, southwest Gansu and northwest Sichuan provinces. The high altitude, low temperature and strong sunlight, and careful selection and feeding by local herdsmen have ensured that the Hequ-line Tibetan Mastiffs are strong and sturdy, with a perfect structure and coordinated movements, wide head and short mouth. Of an aloof, proud and fiery character, they are very loyal to their masters. The Hequ-line Tibetan Mastiffs are mainly tiger-like, and come in diverse colors.

Tibetan Mastiff on the Hongyuan Prairie in the Hequ Region

Tibetan Mastiffs on the Hongyuan Grassland

The Hongyuan Grassland

Pedigree Tibetan Mastiffs can also be found in other pasturing areas in China, for instance, Yunnan's Shangri-la. But since there is only a small number of them and they are distributed sparsely, it is difficult to define them as a separate line.

A view of the Gannan Prairie, the center for breeding Tibetan Mastiffs

# Habits of the Tibetan Mastiff

*Perfect Combination*
*of Ferocity and Ingenuity*

The image of the Tibetan Mastiff as a wild beast that kills people like flies with its knife-edged fangs and drinks blood, as portrayed in some books, is totally exaggerated, and results from people's extreme fear of the animal and ignorance of its temperament. The Tibetan Mastiff is by no means a murderous beast, despite the fact that it retains an appropriate part of its original wild nature. Its

A Tibetan Mastiff, a little Tibetan girl and a lamb. It is hard to imagine from this picture that the Tibetan Mastiff is an extremely fierce animal.

Guarding the livestock of its master is one of the most important duties of the Tibetan Mastiff.

ferocity is merely an expression of its loyalty to its master or a response to a threat. Perhaps the animal is not as friendly and easy to train as other dog breeds, but that is because it is an arrogant breed yearning for freedom, like the untamed horses on the prairie that no coward can subdue or rein

Like other dog breeds, the Tibetan Mastiff is an omnivorous animal. It is actually a misconception to take it for carnivore because of its ferocity. In pastoral areas, the Tibetan Mastiff mainly lives on *tsampa* and milk curd mixed with a little bone soup. *Tsampa* is a Tibetan staple made from *qingke* (highland barley) after being washed, fried and ground into flour. After yak butter is separated from milk by hard churning, the milk is first boiled, and then cooled, and left to become sour. The residue left after filtration is milk curd. Meat is a luxury for the Tibetan Mastiff. However, the smart Mastiff is good at finding special treats for itself by catching small animals, such as marmots, in the grass from time to time. The dog will never touch livestock, even if it is in danger

of starving to death, for it knows that the livestock is the property of its master. As more Tibetan Mastiffs are being introduced to China's inland areas and many foreign countries and regions, their food structure has, of course, become more diversified.

It is a surprising fact that the huge Tibetan Mastiff is not a big eater, but has an appetite even much smaller than that of many other dogs of the same stature. It eats in a leisurely and dignified way. Its noble elegance often arouses awe and respect in people. In fact, Tibetan Mastiff enthusiasts see the dog as their friend, brother or family member, instead of a pet or a tool, which they can dispose of at will.

The Tibetan Mastiff does not bark at random. It will produce a deep roar coming from the chest cavity like muffled thunder only when an intruder approaches the territory under its jurisdiction. Once an intruder enters the dog's area of control, it will instantly take action. The Tibetan Mastiff is born with the attacking skill of beasts of prey, such as lions and tigers—charging at the enemy quietly but at lightning speed, biting the neck, and swaying left and right to kill the enemy. Tibetans in pastoral areas say that one good Tibetan Mastiff can fight off three wolves.

Prancing like a galloping horse and roaring furiously like a lion

The Tibetan Mastiff is an energetic animal. All the time, it roams its territory, except for a nap at noon to recover its strength. One Tibetan Mastiff can guard an area of several sq km with hundreds of head of cattle and sheep and other property on it. When night falls, the Tibetan Mastiff gets on the highest alert for its major enemies—wild beasts, including wolves and leopards—that begin to launch sneak attacks on the inhabited places in the pastoral areas. If one Mastiff encounters a pack of wolves, it will inform other Tibetan Mastiffs and domestic dogs in the area by howling. Tibetan Mastiffs are innately brave fighters, without any trepidation or flinching even in the face of death. Numerous true stories about how they fought bloody battles and saved their masters even at the cost of their own lives are well known all over the pastoral areas.

1.Tibetan wildlife — marmot

2.Tibetan wildlife — wolf

3.Wild donkeys on the grassland

4.Tibetan wildlife — Tibetan Antelope

5.Tibetan wildlife — hare

6.The eagle is revered by Tibetans. In the past, many Tibetans observed the custom of "sky burial," by which they let eagles eat the bodies of the dead so that their souls could be taken to heaven by the birds.

A Tibetan named Sherab Odser living in the Hongyuan Pastoral Area in northern Sichuan Province told us a true story about his two Tibetan Mastiffs. On a snowy night, Odser put all his sheep in the fold beside his tent. At midnight, he heard a noise that sounded like a wolf attack. He rushed out of the tent, only to find that all the sheep and his two guardian Tibetan Mastiffs had disappeared, leaving the corpse of one sheep lying

Fine horses and Tibetan Mastiffs are important companions of Tibetan herdsmen.

outside. The violence of the snowstorm forced Odser to wait until morning before investigating further. The next day, he found each of his Tibetan Mastiffs, covered in blood, guarding one half of his remaining sheep. The carcasses of four wolves lay nearby.

1. Drolma and her ornaments

2. Drolma and her Tibetan Mastiff

3. An old herdsman and a Tibetan Mastiff

4. Children in pastoral areas

5. A Tibetan Mastiff and its young master

6. The Tibetan prayer wheel is often spun as an expression of religious piety and a prayer for good luck.

The Tibetan Mastiff is renowned for its independent thinking and judgment ability, as well as noble qualities such as outstanding adaptability, remarkable memory, composure and courage. In addition to obeying simple orders, the Tibetan Mastiff can make judgments according to its observation of the daily words and actions of its master and other people. For example, when a Tibetan Mastiff finds itself in a new environment, it never snarls or bites, or listens to any stranger's order, or fawns on anyone who gives it food as other dog breeds do. On the contrary, the Tibetan Mastiff will calmly and quietly observe the environment and people around to discern the family members, friends, territory, and properties of its new home. Only after it makes a judgment will the dog integrate into the new home and fulfill its duties. It is astonishing to learn that the Tibetan Mastiff can distinguish the status of each family member. The most powerful person, namely

A herding family and their Tibetan Mastiff

A Tibetan Mastiff and her puppy

Two young pals

the head of a household, mainly refers to the male host in China's family structure. The children of the hosts are the dog's important friends, ranking second in the obedience sequence. The female host takes the third place, and people hired by the hosts are in the fourth place. As regards the hosts' friends, the Mastiff shows friendliness, but maintains its dignity, pays attention, but keeps its distance. The dog seems to be fully aware that, as the personal guard of its master, it does not have to serve or obey any guests of the family.

The Tibetan Mastiff breeds only once a year, like many wild animals, and the female gives birth to three to nine puppies at a time. The female Tibetan Mastiff bears the significant responsibility of feeding and training young Mastiffs. One month-old Mastiffs begin to receive their mother's strict and patient training in charging, biting and fighting skills. Having been tamed and raised by people for several thousand years, domesticated dogs no longer need to allay their hunger by preying on other animals. Tibetan Mastiffs, however, amazingly, pass on this primitive instinct to ensure that their cubs can live independently after leaving their mother.

བོད་ཁྲི།

# Present Situation and Future
# of the Tibetan Mastiff

Western China has been a relatively remote area since antiquity, due to severe weather conditions and harsh terrain. The majestic snow-capped mountains and emerald-green grassland from Qilian Mountains to the Yarlung Zangbo River actually contain countless treasures, one of which is the Tibetan Mastiff.

As the mystery of the Tibetan Mastiff is gradually unveiled, more and more people are becoming interested in it and doing research into it. However, there is a danger that this species may become extinct.

Fortunately, many people of broad vision have perceived this problem, and are attempting to protect the species through pedigree breeding. As China's economy advances, the number of individuals, including entertainment and sports stars, artists, and even common people, who love the Tibetan Mastiff and can afford to raise them is on the rise. The China Tibetan Mastiff Association, chaired by the famous Chinese long-distance running coach Ma Junren, has been established, and bases for research on the breeding of Tibetan Mastiffs are springing up all over China. In addition, the mass media, with the Tibetan Mastiff Union website as the spearhead, spare no effort to promote knowledge about the Tibetan Mastiffs and its protection.

# Appendices:

## 1. How to Get Along with a Tibetan Mastiff

1) Points for attention when first meeting a Tibetan Mastiff

The Tibetan Mastiff is not as dreadful as described in legends. You can feast your eyes on the rare dog as long as you heed the following points:

First, you have to greet the owner before you meet the Tibetan Mastiff, and then approach it in the company of its master. The Tibetan Mastiff's steadfast loyalty is widely known. At the same time, it also has a strong sense of territory, which is a common trait among dogs. Rash intrusion into its territory without the accompaniment of its master may bring about disastrous consequences.

Second, although the Tibetan Mastiff's eyes can hardly discern colors, experts who have bred Tibetan Mastiffs for years warn that the dog has a fierce response to bright yellow and scarlet. Therefore, it is advisable not to wear clothes in these two colors in front of a Tibetan Mastiff.

Third, never try to get close to a Tibetan Mastiff, because it is impossible to become intimate with the dog at the first meeting. After all, the Tibetan Mastiff is not a pet dog. Many first-class Tibetan Mastiffs permit no one other than their master to approach them. Even their feeders are kept at a distance, not to mention strangers. Moreover, you should pay attention to the expression of the dog. Generally speaking, all Tibetan Mastiffs will be on guard against the appearance of a stranger by fixing their eyes upon the stranger, barking continuously, and getting ready to charge and bite. If you see a Tibetan Mastiff bristling and assuming a hostile look, with the eyes suddenly radiant and fierce—signs of a probable attack—you must be very careful.

However, experts say that the Tibetan Mastiff's high degree of intelligence enable it to quickly recognize the guests and friends brought in by its master, and it will then be well-behaved. The dog can also discern people harboring ill will toward its master, and then it will show its hostility.

The Tibetan Mastiff will catch and eat small animals from time to time, but it will never touch the cattle or sheep of its master.

2) How to make friends with a
   Tibetan Mastiff

It is indeed impossible to establish a close relationship with a Tibetan Mastiff at the first meeting, and its ferocity makes you feel that it is extremely difficult even to approach it. But if you have enough courage and patience, you can finally make friends with it.

Step one: When you approach the dog, it will bark at you out of its instinctive sense of territory. Any betrayal of timidity will make following contacts more difficult. Neither should you glare at it with hostility

Once an intruder enters the territory of a Tibetan Mastiff, the dog will take action without hesitation.

or scold it loudly, which will only make the Tibetan Mastiff regard you as an enemy. The best way is to pretend to be deaf and dumb, and smile at it or find other things to do in front of it, regardless of its snarls. After a while, the dog will quieten down, and begin to observe the situation.

An increasing number of Tibetan Mastiff breeders are emerging even in China's cities.

Step two: Approaching a Tibetan Mastiff.

a. Ensure that it always has drinking water, which should be refreshed regularly.

b. When the dog is hungry, give it only a little food, and leave immediately after it finishes eating. Do this every two hours. When its will and strength have been worn down to the minimum by hunger, your approach or even light stroking will not cause strong resistance.

c. Feed the dog regularly with the normal amount of food and talk with it often.

Step three: Establishing your authority as the master. There is an old saying in China: "Kill the chicken to frighten the monkeys," namely punish someone to warn others. You can draw on this method by scolding and punishing some other animals in the family to send the Tibetan Mastiff the message that you are the head of the family, and thus establish your authoritative status.

A herding family and their Tibetan Mastiff

# China's Tibetan Mastiff

བོད་ཁྱི།

Step four: Now you can take your beloved dog out for a walk from time to time. As a freedom-lover, the Tibetan Mastiff will become nervous and violent if it is fastened up or caged for long periods. Frequent outdoor walks can relieve its fretting, and help build up affection between you and the dog.

If you complete the above four steps, congratulations! You have truly become the master of a Tibetan Mastiff. If you are not successful in carrying out those steps, I should send you even warmer congratulations, because an untamed Tibetan Mastiff is as invaluable as an untamed horse. This kind of Tibetan Mastiff cannot be completely tamed by ordinary means.

You must be rigorous toward it. When it rushes at you to try to bite you, you must resolutely lash hard at its mouth with a strap or tree twig to make it feel pain. (Do not hit other vital parts of the dog.) By all means, avoid being soft-handed or merciful. You should scold it loudly while lashing it. Immediately after that, stop feeding the dog for a whole day, and do not comfort it out of pity. If the Tibetan Mastiff still does not obey you, repeat the process until it starts restraining its temper. Then you can treat it with a normal tone and approach.

## 2. Dog Breeds with a Tibetan Origin

Many kinds of native Tibetan dog breeds, such as Tibetan Terrier, Lhasa Apso and Tibetan Spaniel, all possess an ancient mysterious oriental disposition and are associated with a deep religious significance. Closely related in blood, they have a lot of similar characteristics in appearance and structure.

| Breed | Tibetan Terrier |
|---|---|
| History | The Tibetan Terrier originated in Tibet, China, about 2,000 years ago. It was chiefly raised in monasteries, and was considered a "Luck Bringer" or "Holy Dog" by the Tibetans as a whole. The breed was introduced to India in 1920, and to the UK in 1930. It was officially recognized in 1937 and 1973 in the UK and the US, respectively, and is now found all over the world. |
| Temperament and Function | Intelligent, good-natured, loyal, easy to train, and on guard against the appearance of strangers. It is a perfect companion and also a good guard dog. |
| Appearance | Drooping ears set fairly high on the skull, short and strong muzzle, black eye rims; height equal to length, straight and short forelegs, relatively longer hind legs; double coat in white, black or gold colors, or a mixture of two or three of the above colors |
| Height and Weight | 30-40 cm; 8-14 kg. |

| Breed | Lhasa Apso |
|---|---|
| History | The Lhasa Apso's history dates back at least 2,000 years. It was first kept by lamas and nobles as a pet, and later as a monastery guard dog. The dog was also regarded as sacred, and it is said that the soul of its master after death would be attached to the dog and bring good luck to the dog's master. |
| Temperament and Function | Sturdy, genial, determined, and on guard against the appearance of strangers. It is an excellent family pet. |
| Appearance | Narrow skull, very long whiskers; length longer than height; well-developed hind legs; coat hair separated at the spine toward left and right in honey, golden, dark-gray, or tawny colors |
| Height and Weight | Male, 25 cm; female, a little smaller; 4-6 kg. |

| Breed | Tibetan Spaniel |
|---|---|
| History | The Tibetan Spaniel originated in Tibet, China. Although it is called a hunting dog in Chinese, it has never actually been one. It is said that the dog was trained to help lamas to rotate their prayer wheels. Tibetan Spaniel's average life expectancy is 13 to 14 years. |
| Temperament and Function | Independent and self confident. The Tibetan Spaniel was Lamas' companion in ancient times, now the pet of more and more common people. |
| Appearance | Comparatively long muzzle and legs, milky, creamy, brown or dark-brown coat |
| Height and Weight | 24.5-25.5 cm; 4-7 kg. |

**图书在版编目（CIP）数据**

中国藏獒／黎靖，任刚主编.－北京：外文出版社，2005
ISBN 7-119-04272-6
I.中...  II.①黎...②任...  III.犬－基本知识－中国－英文
IV.S829.2

中国版本图书馆 CIP 数据核字（2005）第 115939 号
（鸣谢：北京得宠天下宠物文化发展有限公司）

## 中国藏獒

| | |
|---|---|
| 策　　划： | 肖晓明 |
| 责任编辑： | 杨春燕　刘芳念 |
| 英文翻译： | 韩清月　李　洋 |
| 英文审定： | Paul White　贺　军 |
| 撰　　稿： | 黎　靖 |
| 摄　　影： | 任　刚 |
| 版式设计： | 精诚设计 |
| 印刷监制： | 韩少乙 |

© 2006　外文出版社
**出版发行：** 外文出版社（中国北京百万庄大街 24 号）
邮政编码 100037　http://www.flp.com.cn
印　　制： 北京市大容彩色印刷有限公司
2006 年第 1 版第 1 次印刷
（英）
ISBN 7-119-04272-6
04800
85-E-601P